Human Impact

Human Impact

OUR RELATIONSHIP WITH CLIMATE, THE ENVIRONMENT, AND BIODIVERSITY

Edited by Kate Stone and Shayna Keyles

Science Connected

Human Impact: Our Current and Future Relationship with Climate, the Environment, and Biodiversity
Published by Science Connected, Inc.

Editor in chief: Kate Stone
Managing editor: Shayna Keyles
Lead copy editor: Jess Romaine
Copy editors: Małgorzata Pachol, Rachel McCabe, Helen Cheng
Cover photographer: Max Goldberg

© 2019 by Science Connected
180 Steuart Street #190213
San Francisco, CA 94119 USA
www.scienceconnected.org

All rights reserved. No part of this book may be reprinted or reproduced or utilized in any form or by any electronic, mechanical, or other means, now known or hereafter invented, including photocopying and recording, or in any information storage or retrieval system, without permission in writing from the publisher.

Some contents of this book have been previously published in Science Connected Magazine in 2018 and 2019. They have been modified and expanded upon for inclusion in this volume.

This book is independently authored and published, and no sponsorship or endorsement of this book by, and no affiliation with, any trademarked product mentioned or pictured within is claimed or suggested. All trademarks that appear in the text in this book belong to their respective owners and are used here for informational purposes only.

This book mentions the Next Generation Science Standards, a registered trademark of Achieve. Neither Achieve nor the lead states and partners that developed the Next Generation Science Standards were involved in the production of this product, and do not endorse it.

Contents

Introduction	7
Artificial Night-Lights Are Growing, Getting Brighter	11
Globalization and Its Environmental Impact	17
Imagining Future Wastewater Solutions	23
The Impact of Developing Biofuels on Travel Emissions	27
Why Plastics Are Dangerous to Our Health	33
Plastic Pollution: An Emerging Threat Beneath Our Feet	39
Microbes Help Plants Survive Heavy Metal Stress	45
Does Habitat Fragmentation Affect Biodiversity?	49
An Evolutionary Approach to Conserving Plant Habitats	53
Do Humans Influence Jellyfish Coastal Populations?	57
Is Climate Change Causing These Moose to Shrink?	61
Plants Can Win a Battle against Aphids	67
Do We Really Need Fertilizers To Grow Crops?	71
Forest Restoration, Not Plantations, Will Curb Global Warming	75
Commodity-Driven Deforestation Threatens Forests	81
Living with Wildfires: Fighting Fire with Fire	87
How Wildfires Start Their Own Weather	91

Questions for Discussion	95
Citizen Science Resources	103
References	111
About the Authors	123
About Science Connected	127
More from Science Connected	129

Introduction

Kate Stone and Shayna Keyles

All of us who are paying attention know the earth is changing, and we know that we're part of the reason. We watch as average temperatures increase, severe weather intensifies, populations dwindle and disappear, and diseases begin to appear again. And we know that if we want to fix things, we must examine our actions and do better.

This book examines our changing world in 17 articles written by science communicators for Science Connected Magazine. These articles don't shy away from the harsh truths we're currently facing; we're seeing more wildfires, more pollution, and more pests, for example. But this isn't doom and gloom reporting; this is a glance at the future, at a way we can repair some of the damage that's been done.

For example, the Adyar River in Chennai, India, once supported a thriving culture, economy, and natural habitat. Then it was filled with untreated wastewater. Next, it became a landfill. The resulting sludge, thick with debris, refuse, and human waste, was hazardous to touch and no longer recognizable as a river. It seemed like a lost cause to some, but in 2012 the local citizens started a petition that persuaded the government to address the issue, leading to a large-scale restoration project. This shows that informed and empowered people can be a powerful engine for change, and the hard work of dedicated researchers and scientists can pave the way to innovative solutions.

With this article and others, this book puts the latest research findings on climate, environment, and biodiversity into your hands. How have humans impacted the environment and how can we adjust that impact going forward?

How to Use This Book

In the following pages, you'll read about light pollution, wastewater treatment, plastics, habitat fragmentation, extreme population changes, pests and fertilizers, problems facing our forests, and wildfire management. Some articles detail the problems and others suggest solutions. Taken as a whole, this anthology provides an eagle-eye view of some of the ways humans are changing our planet, for good and for bad.

The book includes discussion questions to use in a classroom or community setting to help readers synthesize and contextualize the information.

Educators whose classes are aligned with the Next Generation Science Standards (NGSS) may find this book useful in addressing the following individual standards and their associated performance expectations:

HS-LS4-2 Biological Evolution: Unity and Diversity

Mastery of this standard allows students to construct an evidence-based explanation of evolution based on the following factors:

- the potential for a species to increase in number
- genetic variation in individuals of a species as a result of mutation and sexual reproduction
- competition for resources
- survival of the fittest

HS-LS2-7 Ecosystems: Interactions, Energy, and Dynamics

Mastery of this standard allows students to design, evaluate, and refine a solution for reducing the impacts of human activities on the environment and biodiversity. Examples of human activities may include urbanization, damming waterways, and disseminating invasive species.

The Next Generation Science Standards were developed by educators, content experts, and policymakers, using as a guiding document the Framework for K-12 Science Education from the US National Research Council. For the full text of these standards, as well as the accompanying performance expectations, please visit http://nextgenscience.org.

For readers who are interested in making a positive impact, we also include a guide to citizen science projects related to the issues discussed in this book. Citizen science, sometimes called community science, is a collaborative scientific process in which nonscientists and scientists work together on a project. It can be contributive, in which participants gather data; collaborative, in which participants may also analyze or interpret the data; or co-created, in which participants are involved in all levels of a project. Our guide has been curated by Caroline Nickerson from SciStarter, an important community and database of over 2,700 ongoing citizen science projects.

We also offer classroom discussion guides and lab manuals that further explore some of the issues raised in these pages. For more information about these resources, see the "More from Science Connected" section of this book or visit www.scienceconnected.org.

Artificial Night-Lights Are Growing, Getting Brighter

Neha Jain

As soon as it gets dark, street lights, which have become widespread in the developed world—and are rapidly expanding in the developing world—are switched on. Indeed, since the second half of the twentieth century, Earth has become brighter at night. Now, new satellite-based research shows that our outdoor artificial night-lights are still spreading to more areas on Earth and have gotten brighter over the past few years.

How Night-Lights Impact the World

Outdoor lighting is regarded as a necessity, especially in highly populated areas, but artificial night-lights are an environmental pollutant, disrupting processes such as light-dark cycles in nocturnal animals—those that are active at night—and affecting plants and microorganisms. Prolonged exposure to artificial night-lights has also been suspected of affecting our health and sleep patterns.

During the second half of the twentieth century, outdoor lighting grew at a rate of 3 to 6 percent per year. The artificial sky glow created by our night-lights is one of the most visible forms of light pollution. Most of the populations of Europe and the United States are exposed to light-polluted skies at night. As a result, almost two-thirds of Europe and 80 percent of the United States cannot see the Milky Way.

Even "pristine" or protected areas that lack any human activity are affected by lights hundreds of kilometers away. Consequently, pristine skies that are not shrouded by sky glow are becoming rare in many regions.

A New Way to Research Night-Lights

Christopher Kyba, a researcher from the GFZ German Research Centre for Geosciences in Potsdam, Germany, wanted to find out whether the use of night-lights is still on the rise globally or has leveled off. "I don't think anyone has really looked at global lighting change before, at least not with the focus on whether lights are increasing," he says.

His team used a calibrated satellite radiometer sensor to measure global changes in the scale of night-light emissions as well as the brightness or radiance of lights from 2012 to 2016. This sensor measures nighttime radiance from light between 500 to 900 nanometers (nm) in wavelength (human visual system is 400-700 nm) at a finer spatial resolution of 750 meters, which can detect changes in lighting at the neighborhood level.

"It's brightening most rapidly in developing countries, but it's also brightening in wealthy countries that were already bright in 2012 when the study began," says Kyba. "Our results are consistent with the hypothesis that increasing 'energy efficiency' of LEDs is not saving energy at the national and global scales, but instead is simply resulting in greater light use."

More Night-Lights Lead to "Loss of Night"

The results of this study showed that from 2012 to 2016, Earth's artificially lit outdoor area rose 2.2 percent per year. Areas that were continuously lit grew brighter by 2.2 percent. Most countries showed an increase in brightness, particularly those in Africa, Asia, and South America. Brightness remained stable in some of the world's brightest countries such as the United States, Spain, Italy, and the Netherlands. Only a few regions showed a drop in brightness, notably war-torn nations such as Syria and Yemen.

The main aim of the LED "lighting revolution," which involves transitioning to the more energy-efficient light-emitting diodes (commonly known as LED lights) as opposed to the more energy-consuming incandescent and fluorescent lights, is to decrease overall energy consumption.

As cities switch from orange sodium lamps to more energy-efficient "white" LED night-lights, the satellite sees this as an overall decrease in light because, as Kyba explains, it cannot see the blue light emitted by LED lamps, which is below 500 nm in wavelength. For this reason, Kyba's team expected to see overall decreases in light in countries such as the United States and Germany. But, instead, they saw that total light in Germany actually increased, while it was constant in the United States.

"This means that while many cities appeared to get darker, many other places must have installed new or brighter lamps," Kyba reports. "These new lamps need energy, and therefore the savings at the national level aren't as large as what you would think, based on all the reports of LEDs saving energy."

In an effort to promote the sustainable development of cities, governments strive to reduce the cost of LED night-lights. However, falling costs may translate to more lights installed, resulting in more energy consumed overall.

"Cheaper light from LEDs is likely to lead to new uses of light outdoors. For example, in the past it was not common to light bridges, but this is becoming increasingly common," notes Kyba. He adds that this can harm the fish that swim beneath.

There are many cases of street lights installed in locations where they are not necessary or just too bright to serve the purpose. Kyba compares the ring highway around Berlin, which is unlit and has no speed limit in some places, with the ring road around Calgary, which has a speed limit of 110 kmh (68 mph) and is lit. "If you can drive 170 kmh on an unlit road in Berlin," he asks, "why does the ring road around Calgary with a 110 kmh limit need to be lit?" Another example is in the southern Indian city of Thiruvananthapuram, where costly high-mast lamps have been installed at unsuitable locations, blinding drivers with glaring light and generating high power bills.

While Kyba agrees that we should have lighting development, he asserts that "it should be done with effective lighting" and that "it would be a terrible shame if the lights that were installed were ugly, glaring, and didn't necessarily improve the visual situation."

Download the citizen science app "Loss of Night" on Android and iOS to monitor sky glow by simply counting stars.

For more examples of good and bad lighting, head over to Kyba's blog at lossofthenight.blogspot.com.

If you are interested in the study and want more information about effective lighting, check out the International Dark-Sky Association.

This study was published in the journal *Science Advances*.

Globalization and Its Environmental Impact

Megan Nichols

Broadly speaking, globalization refers to the increased interdependence of nations and the way people from different cultures and geographic locations can receive goods or communicate with each other thanks to free trade and information technology, among other things. But it's a much more complex phenomenon than that, and it's necessary to have an all-encompassing understanding of what it entails. Here, we'll examine how globalization affects the environment in both positive and negative ways, and what changes could be made to ease its adverse effects in a post-globalization world.

Globalization Causes Environmental Changes that Affect Health

In his in-depth academic article, Anthony J. McMichael from the National Centre for Epidemiology and Population Health at the Australian National University asserts that the effects of globalization are not separate, encapsulated events, but multifaceted phenomena that have a domino-like effect on the environment. He discusses how climate change will make it exceptionally difficult to grow crops, contributing to food scarcity crises, and he predicts that globalization will be partially to blame for the introduction of new diseases to particular regions. Yet another consequence that McMichael considers, but which doesn't often appear in other discussions about globalization effects, is the changes in ocean waters and how they cause lower protein content in fish.

Climate change can affect agriculture in various ways; more or less rainfall, temperature extremes, introduction of new pests, and changes in carbon dioxide and nitrogen levels could all have consequences for crop yields.

Although cons of globalization are numerous, there are some positives too. For example, milder winters could reduce the death rate from health issues such as stroke. It may also become more difficult for mosquitoes to survive as the weather becomes hotter and drier due to climate change, which could lead to a decline in fatalities resulting from mosquito-borne illnesses such as malaria.

According to McMichael, globalization has already caused health-related changes, and still more changes will occur at an especially rapid rate if numerous parties don't intervene promptly. More specifically, the health sector must collaborate with other industries and figure out how to alter the ways in which humans produce and share things and engage in activities that comprise their existence while keeping climate change and its effects in mind.

An American Professor, a Haitian Business Leader, and a Common Goal

One positive effect of globalization as it relates to the environment is that it facilitates the coming together of people from different backgrounds to tackle environmental issues. Professor Stephen Blair Hedges, director of the Center for Biodiversity at Temple University, and Philippe Bayard, CEO of Sunrise Airways and president of Société Audubon Haiti, a leading conservation group, worked together to establish Haiti's first private nature reserve. The protected areas of the reserve act as havens for species most at risk of extinction, a particularly dire threat given the connection made between forest cover and biodiversity, and the fact that—as Hedges points out—Haiti is one of the most deforested countries in the world. The problem at hand is severe, but the project shared by Hedges and Bayard illustrates how globalization helps people defy geographic boundaries and pool their knowledge.

Hedges has also coauthored research of which one of the primary goals is to improve forest cover reporting data and connect the findings to biodiversity. According to Hedges, the current method used by the United Nations to monitor the world's forests is not sufficiently detailed. However, the efforts to preserve Haiti's disappearing species could lead to positive changes at the governmental level and yield similar collaborations between experts.

Sustainability As Part of Corporate and National Strategies

Although individuals can take personal measures to support environmental sustainability, Ravi Fernando, operations director of the Malaysia Blue Oceans Strategy Institute, argues that helping the planet thrive should be part of the corporate and national strategies devised by people in leadership roles, because they are the ones who have the power and resources necessary to spur positive developments.

In his paper "Sustainable Globalization and Implications for Strategic Corporate and National Sustainability," Fernando points out how climate and poverty are "the two all-pervading global sustainability challenges." And even if leaders think about these challenges, it's typically only with a short-term mind-set. But when visionary leaders work together with consumers who keep abreast of relevant developments and maintain real-world perceptions, changes do happen. National policies are a good starting point, but only when supplemented with sustainable business practices and the support of consumers willing to do things differently than before in favor of sustainability.

Fernando also talks about what tremendous influencers the most powerful nations and businesses are. As such, they have a substantial responsibility and must remember that when they take action to either help or harm the environment, the world watches, and some people will respond in kind. Whether independently or in collaboration with others, corporations and governments can combine knowledge and other resources to make gains that will benefit the planet.

As for individual efforts, people could help by reacting favorably when companies and governments make choices that are advantageous to the environment. They could also voice their disapproval of any proposed legislation that could potentially harm the environment. And on a personal level, they could start changing their ways, for example by purchasing goods from more sustainable companies. Because every effort is worth it.

Imagining Future Wastewater Solutions

Emily Folk

Wastewater represents a serious risk to human health in both developing and developed countries. Through industrial, commercial, agricultural, and domestic activities, affected sources of water cause illness, disease, and even death. One particular case study serves as an example of these dangers.

The Adyar River in Chennai, India, once supported the area's economy and culture. With the introduction of untreated wastewater, the river soon became an active landfill, inundated with debris and refuse, with a thick consistency that no longer allowed for safe interaction (Hariram, 2017). In 2012, a citizen-driven petition persuaded the Tamil Nadu government to address the issue, leading to a large-scale restoration plan to reduce pollution in the Adyar River. With the program currently in its third phase, rehabilitation continues, and the project has seen considerable success. We'll examine the details of that project and explore other solutions that promise to advance wastewater treatment and prevent issues like those in Chennai.

Restoration Efforts in Chennai

Restoration in Chennai began with the construction of over 300 sewage treatment plants. The government's investment in this infrastructure ensured fewer instances of pollution, laying the foundation for future progress. These treatment plants would function as a preventive measure rather than as reactive damage control. Phase two involved eco-restoration, planting mangroves along the estuary stretch of the river. The third phase, presently underway, entails continued ecological restoration with the desilting of riverbeds. This model of nature-based solutions and gray infrastructure has applications in other affected areas around the world.

Projects like those in Chennai will prove important to replicate, as up to 80 percent of wastewater enters waterways untreated. In disadvantaged countries where sanitation and hygiene are challenging to maintain, local governments can study the Adyar River and use it as a framework to structure their own programs. The restoration project in Chennai is only one example of modern wastewater treatment. Beyond its model, other solutions have just as much potential.

Innovative Solutions for Treatment

Various methods of environmental remediation have proven effective in the treatment of wastewater. One of these methods is bioremediation, which includes the use of microorganisms or other forms of life to eliminate environmental pollutants and detoxify an affected area. Multiple approaches to bioremediation exist. Phytoremediation is a type of bioremediation that uses plant life; this is often proposed to counter the bioaccumulation of metals (Phillips, 2019).

Beyond bioremediation, other kinds of environmental remediation have been applied in wastewater treatment. Nanoremediation is another strategy with considerable promise. This is a comparatively new technology that uses nano-sized particles for the remediation of polluted water and soil. The implementation of zero-valent iron has successfully treated acidic water carrying heavy-metal pollutants, with the ability to neutralize and immobilize contaminants (Das, 2018).

Reverse osmosis is also a reliable solution for producing drinking water during natural disasters such as floods. The process begins when water from an outside source meets a semipermeable reverse osmosis membrane, filtering out the contaminants. The reject stream removes dissolved solids while permeating water passes through the other side of the membrane, safe for consumption (Herold, 2018).

A Secure Environment for Future Generations

Wastewater represents a serious risk to human health, but the dedication of researchers and scientists has given way to innovative solutions. The large-scale restoration project in Chennai is only one example of their ingenuity, integrating nature to impressive effect. Bioremediation, nanoremediation, and reverse osmosis are no less deserving of attention, each with its own distinct advantages. As we move toward the future, wastewater will continue to pose problems in countries around the world. Even so, we possess technologies to address them, with the means to create a secure environment for generations to come.

The Impact of Developing Biofuels on Travel Emissions

Emily Folk

Approximately 29 percent of the energy the United States consumes is for transportation, and about 92 percent of US transportation is fueled by petroleum products. As a result, transportation is the sector that produces the most greenhouse gas emissions in the United States.

With the threat of climate change and other environmental challenges looming, we need to find a way to reduce emissions from the transportation sector. Biofuels are one potential solution.

What Are Biofuels?

Biofuel is fuel made from organic matter known as biomass. This organic matter includes substances such as plant matter, fungi, algae, and animal waste. Various processes exist for making biofuels, but they typically involve using chemical reactions, heat, and fermentation to break down molecules in the organic matter. The resulting products can then be refined to create a fuel that can be used for transportation.

Biofuel is a renewable resource because it can be replenished quickly, unlike fossil fuels. Many biofuels release fewer emissions than fossil fuels, but not all biofuels are necessarily low emission.

Current Biofuel Uses

The earliest automobiles were fueled by biofuels, but petroleum products overtook them after the discovery of cheap oil reserves. Today, most of the gasoline used for cars in the US is blended with ethanol, a type of biofuel. Some cars can also run on just ethanol. The use of biofuels for transportation is more common in Europe where palm oil, a diesel-like biofuel, is widely available.

Other transportation industries, such as aviation, are also starting to explore the use of biofuels. In 2016, United Airlines started using biofuels for regularly scheduled flights, becoming the first airline to do so. United is also partnering with a company called Fulcrum BioEnergy that produces biofuel using waste and household trash. The aviation industry produces 12 percent of CO2 emissions, so switching to cleaner-burning biofuels could have a substantial impact on the health of our environment.

Biofuels and Emissions

So, how do biofuels and petroleum products compare when it comes to emissions?

One study that compared diesel with a soy-based biodiesel alternative found that the soy-based fuel produced 93 percent more energy and reduced greenhouse gas emissions by 41 percent. Evaluating the burning of biofuels versus fossil fuels often shows similar results.

When looking at the wider picture, though, it gets a bit more complicated. Biofuels have other environmental impacts too, namely those involved in growing crops, including the use of land and inputs such as fertilizers, pesticides, water, and fuel used to power the processing and transportation of biofuel products.

A paper published in the journal *Atmospheric Chemistry and Physics*, for instance, argues that agricultural activities related to biodiesel increase N2O emissions. N2O has a stronger warming potential than CO2, and the increased N2O emissions may offset the reduced CO2.

Understanding the full picture of biodiesel and emissions requires looking at all stages of the product's life, a process called life cycle analysis. When the US Department of Energy's Argonne National Laboratory performed such an analysis, they found that emissions from biodiesel were 74 percent lower than emissions from petroleum diesel.

However, the World Resources Institute has calculated that providing 10 percent of the liquid transportation fuel we'll use in 2050 would require about 30 percent of the energy present in our annual production of crops. This suggests that it might not be realistic to use biofuels for all of our transportation needs.

The Royal Academy of Engineering in the United Kingdom has recommended that, to get around these challenges, we should use more waste to produce biofuels, rather than growing crops to use as fuels.

The Future of Biofuels

Biofuel use is growing, but it still makes up a relatively small part of fuel used for transportation. The global biofuels market was worth $168 billion in 2016 and is forecasted to grow to $218.7 billion in 2022.

In the future, we'll likely see new forms of biofuel in use. Today, we mostly use what' are called first-generation biofuels, which are made from sugars, oils, starches, or animal fats. These fuels include ethanol, biodiesel, and methane. Researchers and businesses are now starting to explore more second-generation biofuels, which come from non-food crops, such as switchgrass, willow, and wood chips, as well as agricultural waste. These fuels contain more cellulose, the substance that makes up plant cell walls, which could produce biofuels that are more efficient and cleaner burning than those we use today.

Further into the future, we may start using third-generation biofuels, which are those produced using algae. Fourth-generation biofuels come from plants or other biomass materials that are engineered specifically to have higher energy yields, lower emissions, and other advantages.

Biofuels may not be the cure-all for our transportation emissions problem, but they will likely play an important role in addressing the issue in the years to come. From petroleum to corn to algae, the way we power our transportation is changing.

Why Plastics Are Dangerous to Our Health

Emily Folk

It is almost impossible to avoid plastics today. They're used to package our food, hold the water we drink, and even print our receipts at the grocery store. Though plastics have made life more convenient than ever before, could there be a downside to their prevalence in our world? Scientists are trying to find out. One area of research focuses on how regular contact with plastic might harm the human body, especially because people unknowingly ingest plastic practically every day.

Small plastic particles most commonly enter the body through the food we eat. When people consume food or beverages packaged in plastic, there's a good chance that some of that plastic may have transferred over to the food or drink itself. The amount of plastic that transfers depends on multiple factors, including the age of the plastic, the temperature of the container, and the nature of the contents. For example, more plastic is transferred when it is heated or when the food stored inside is fatty. Although some substances are more prone to plastic transfer than others, a study by Orb Media (2018) estimated that, on average, more than 90 percent of water packaged in disposable plastic bottles contained plastic particles.

Since it is difficult to avoid exposure to plastic in the industrialized world, scientists have realized that it's essential to understand how plastic affects our bodies.

Is BPA Dangerous?

As many already know, the science doesn't look good for the effects of ingesting plastics and particles absorbed through skin contact. In particular, some scientists are worried about one chemical called Bisphenol A (BPA), which is manufactured and used in polycarbonate plastics and epoxy resins, such as those used to coat metal cans.

One study of 1,455 American adults found that higher concentrations of BPA in urine were associated with higher incidence of diabetes, heart disease, and liver toxicity (Lang et al., 2008), and these results were replicated in a later study by Vandenberg et al. (2010). The cause of this correlation is yet to be determined, however. Here are some of the possibilities: BPA exposure leads to higher risk of these diseases; these diseases cause higher retention of BPA in the body; BPA and these diseases could both be products of a common origin, such as unhealthy diets, which are much more likely to include plastic-packaged processed foods.

In addition to the clear relation between BPA and conditions including diabetes and heart disease, scientists are also concerned about other properties of BPA. It is an endocrine disruptor, which means it can interfere with the functions of the endocrine system, including regulation of growth, development, and metabolism. It has been suggested that endocrine-disrupting chemicals such as BPA could lead to increased risk of cancer, recurrent miscarriages, and fertility issues, all of which add to growing concern in the science community about plastics usage (Carvalho and Marques-Pinto, 2013).

How Does BPA Affect the Body?

Endocrine disruptors work by binding to certain hormone receptors, blocking the effects of important hormones or causing the body to produce too much or too little of some hormones. BPA is a xenoestrogen. It mimics the hormone estrogen in the body and is able to bind to estrogen receptors (Vandenberg et al., 2009) as well as possibly thyroid hormone receptors (Zoeller, Bansal, and Parris, 2005). However, the effects of BPA, once bound to these receptors, are less clear and currently under investigation.

In rodent studies, BPA was found to affect some aspects of sexual dimorphism or development of certain differences between sexes. Perhaps more concerning, researchers found that BPA exposure before birth in rats led to higher incidences of cancers starting in the skin or tissue (carcinomas) as well as increased sensitivity to other carcinogens in the mammary glands (Vandenberg et al., 2009). Although rats and humans are very different and it is difficult to make assumptions about human health from animal studies, this research does suggest the possibility that BPA and breast cancer could be related.

This possible correlation has to do with the fact that BPA is a xenoestrogen, and research has found high estrogen exposure to be a risk factor for breast cancer (Breastcancer.org, 2016). Therefore, some scientists worry that exposure to chemicals that mimic the hormone might have the same effect. More research is needed to fully understand the interaction between BPA and the human endocrine system.

What Can You Do to Avoid BPA?

It is difficult to avoid BPA exposure from plastics altogether, but it is important to take extra precautions for certain periods of your life. BPA may pose a greater risk during prenatal development, childhood, and puberty, because hormones play a more important role during those periods. Not only are BPAs harmful for developing children, but they have harmful effects on adults and even pets. This is why you may have seen water bottles, pet feeding dishes, or baby bottles labeled "BPA-free."

If you're interested in protecting yourself and your family, the best thing you can do is to avoid food and drink packaged in plastic as much as possible. It is also a good idea to keep an eye on the research as scientists learn more about what BPA might or might not be doing to our bodies. Staying informed about the research and BPA-free alternatives can go a long way toward a healthier life for you and your loved ones.

Plastic Pollution: An Emerging Threat Beneath Our Feet

Emily Rhode

Tiny plastic particles that can barely be seen by the human eye have made their way from the soil into everyday items we know—from earthworms to honey to the beer that we drink—bringing toxic chemicals with them wherever they go. The saying goes that what we can't see can't hurt us. Yet what if these unseen particles are not only hurting us but also changing the entire course of biological evolution? Researchers in Germany have issued a new warning that these human-generated "microplastics" could potentially be wreaking environmental havoc right under our feet, and that ignoring the problem could have dire consequences.

"We think that there is a reasonable chance that inland waters and agricultural soils represent a larger portion of environmental microplastics compared to oceanic basins," says Dr. Abel de Souza Machado of Freie Universität Berlin and IGB-Berlin. "However, this is based in limited evidence." The scientists have made an urgent call for increased research into the effects of terrestrial microplastics, citing a small but growing body of evidence that they pose an emerging global threat to human and environmental health.

Plastic Straws in the Spotlight

Plastic pollution exploded into the public consciousness in April 2015 when a shockingly graphic video taken by marine biology doctoral student Christine Figgener went viral. The footage depicts an olive ridley sea turtle wincing in pain as scientists remove a full-size plastic drinking straw lodged in its left nostril.

The backlash from the video ballooned into a worldwide movement to ban single-use plastic straws. Film director Linda Booker was inspired to build on the straw-free campaign by creating a documentary film that she aptly named *STRAWS*. The film highlights plastic contamination in the oceans and follows the stories of school children, entrepreneurs, scientists, and artists who are working to solve the problem.

"I wanted to find a way to demonstrate a global problem in a way that was impactful, while also being careful not to overwhelm people. Straws are small and uncomplicated, but when you start informing people about the vast number of them used daily and that they are non-recyclable, it helps start the conversation about consumption and production of plastics. Once people are made aware that billions are used only once daily, worldwide it becomes a 'sticky' topic, and all of a sudden they start to notice them wherever they go to eat or drink and much too often in streets, ditches, and on beaches as litter."

Plastic on the Nanoscale

The plastic litter that ends up on the ground is exposed to water, sunlight, and even soil microbes. Eventually, it starts to break down. At less than 5 millimeters, these microplastics begin to move through the environment much differently than they did when they were part of a plastic bottle. Broken down into even smaller pieces, things start to get strange. At one ten-thousandth of a millimeter, plastics on the nanoscale can have new properties that look nothing like those of the original plastic bottle. And that can be dangerous.

"When you change the size of the plastic to the nanoscale, it changes the properties in ways that we don't understand yet," says Dr. Brian Chapman, a scientist who studies nanoparticles. "Some can be absorbed more readily into the body, even through skin. [They] can give off the toxic components more easily." These toxins may be causing more harm than the pieces of plastic that can be seen.

Microplastics and nanoplastics have been well documented in marine ecosystems, showing up in huge "garbage patches" in the

Pacific, in commercially harvested fish in the North Atlantic, and even in corals in the Great Barrier Reef. But ocean contamination might be just the tip of the plastic pollution iceberg. Some estimates put the amount of plastics that find their way into soils or inland waters at 32 percent of all plastic waste. Plastics from wastewater sludge that is applied to agricultural fields is one potentially large source. Aerosolized plastics from manufacturing plants is another. Despite several alarming studies about their effects on terrestrial environments, microplastics and nanoplastics in soil have gone largely unstudied. "Continental systems have been understudied compared to aquatic because it is visually and methodologically easier to identify this contaminant in aquatic ecosystems," explains Machado.

The Consequences of Plastic Contamination

The completed research is troubling. When exposed to microplastics, earthworms showed changes in size and burrowing behavior in the soil. Microplastic exposure has also caused negative effects on the reproduction and growth of springtails, a common arthropod found in soil. The researchers fear that these are just some of the ways that human-generated plastic contamination could be putting evolutionary stress on organisms.

Plastic particles have been found in the beer we drink and the seafood we eat. When asked if plastics in soils could have more harmful effects on the environment than what we are already seeing in the oceans, Machado said that "it is very likely that not even all possible effects of microplastics in aquatic systems have been proposed. What we can say is that it is within terrestrial and continental systems that most of plastics are first produced, used, and disposed, and therefore they have first the chance to impact biota. Other attempts to compare potential effects in the marine and continental environment by now could be highly conceptual and not based on factual evidence."

By eliminating plastics from our daily lives, experts believe that we can slow their accumulation in the environment. "I hope in addition to straws, people will switch to non-plastic shopping bags, water bottles, carrying their own utensils, etc., and think more about the packaging options of their food and beverage purchases," says Booker. "The next level is the other items in our households (especially kitchens) that used to be glass, metal, and wood and avoiding single-use plastic." But until more scientific evidence is gathered, we are left to wonder what harm these unseen particles are causing, and what future dangers they may pose.

Microbes Help Plants Survive Heavy Metal Stress

Radhika Desikan

When you hear the term *heavy metal*, what do you think of? Music or chemistry? Exposure to heavy metal music can cause stress in some humans. But what about chemical heavy metals? Are they good or bad for the environment and living organisms?

In chemical terms, heavy metals are elements in Earth's crust that have a high density; they include zinc, copper, iron, silver, gold, arsenic, lead, and cobalt, to name a few. While trace amounts of heavy metals such as copper, iron, cobalt, and zinc are essential for many cellular processes, an excess of heavy metals including lead, arsenic, mercury, and cadmium is considered toxic to living organisms and the environment.

However, using heavy metals in our everyday life is unavoidable. Without them, we would not have mobile phones, pipes, railroads and bridges, automobiles, or even certain medicines. Heavy metal contamination of the environment occurs primarily because of human activities such as mining, relying on coal-fired power stations, using pesticides, and producing industrial wastes and vehicle emissions. Deposition of excess heavy metals in the soil impairs its quality, which results in poor plant growth and takes its toll on agricultural land. Because heavy metals persist in the soil for a long time and cleaning contaminated soil with industrial methods is expensive, there is a need for other options.

Plants to the Rescue

For a number of decades, plant scientists have been cleaning up contaminated soil through *phytoremediation*, a process involving the use of plants and their associated microbes to reduce concentrations of contaminants, rendering the soil less toxic and more arable. While sometimes plants are used to absorb heavy metals (a process called *phytoextraction*) and clean the soil, this is not always beneficial, as the plants, having absorbed toxic heavy metals, need to be discarded. An alternative approach that has been studied is to use microbes that are beneficial to plants and that occur naturally in the soil.

Plants Use Microbes to Survive Stress

Endophytic microbes (bacteria or fungi that live inside another organism) and the host plant they reside in develop a symbiotic relationship: the microbes gain nutrients from the plant, while the plant uses the microbes to better extract nutrients from the soil. It is not too clear, however, how exactly this interaction occurs. Whilst it is known that plants produce many growth-regulating hormones, little is known about hormones produced from microbes that affect plant performance in the field. In addition, there is a dearth of information about the interaction between endophytic microbes and crop plants to remove heavy metals from the soil.

Recently, Muhammad Ikram and colleagues have isolated an endophytic fungus from the *halophytic* (salt-tolerant) medicinal plant *Solanum surattense*. They identified the hormone auxin from the fungal endophyte *Penicillium roqueforti* (called strain CGF-1), which had a growth-promoting effect on the inoculated wheat plants. Moreover, CGF-1 enhanced the tolerance of the wheat plants to heavy metal stress, compared to the plants that were not exposed to CGF-1.

The inoculated wheat plants showed an increase in photosynthetic capacity, nutrient uptake, and concentration of compounds indicating stress tolerance. Importantly, the scientists observed that CGF-1 colonized the wheat roots and absorbed the heavy metals, thereby reducing the plant's heavy metal uptake. The fungi were not present in the upper shoot parts of the plant, which restricted the presence of the heavy metals to the roots. The mechanism by which the fungus achieves heavy metal uptake, whilst allowing the host plant to grow better, remains to be discovered. Nevertheless, this study provides a possible sustainable solution to cleaning up contaminated land, while also growing useful crops.

Given that contamination of arable soils with heavy metals is a persistent problem on Earth, it is important that novel solutions such as the one described above be considered for use on a larger scale. Employing an eco-friendly biological approach not only eliminates the chances of further land contamination and protects leafy crops but also readjusts the balance in the ecosystem, protecting the environment that we humans are inadvertently damaging.

Does Habitat Fragmentation Affect Biodiversity?

Jacqueline Salvi de Mattos

Most scientists used to believe that habitat fragmentation was a real threat to biodiversity, but some controversial ideas have made this a potential topic for further discussion. Iconic experiments in the Brazilian Amazon, for example, have shown the edge effects on the ability of small patches to retain species. *Edge effects* relate to the possible outcomes of having too much border on chopped forests, which creates a boundary between two habitats. Where there is deforestation, *patches* are the remaining pieces of natural vegetation among other human-affected areas.

Recent studies, however, discuss our understanding of this relationship between patched forests and biodiversity in a different way, including an article by the editors of the journal *Biological Conservation*.

The New Conservation Science

The latest discussions in the conservation field have revealed many inconsistencies on the fragmentation versus biodiversity issue. *Fragmentation* means discontinuities in natural environments, mainly created by human activities, that affect populations and the ecosystem in many ways. *Biodiversity* refers to the huge variability of life on the planet. Understanding the inconsistencies in this research is very important in choosing which kind of habitat to prioritize for conservation practices. Some researchers have recently pointed out that those inconsistencies might be due to the confounding effect of spatial scale, habitat size, and fragmentation. In addition, fragmentation has not been statistically significant in many studies as a potential driver of biodiversity loss, showing that the method of maintaining few larger areas for conservation is less effective than maintaining many small patches within managed areas.

On the other hand, there are studies that have produced opposite insights into the fragmentation idea, which are that fragmentation has a lot to do with habitat loss and should therefore be considered a problem. One of the effects of fragmentation in biodiversity, in these cases, is increasing the abundance of disturbance-adapted species, which are species that can adapt well and are therefore able to persist in the disturbed habitats, as well as invasive species, which are not endemic (native) to these habitats and can harm species of conservation concern.

The Zombie Idea

Scientists are already familiar with the fragmentation issue on a more restrictive, patchy scale. But the argument here is regarding broader, more inclusive scales, which might represent a gap in the conservation research. There are many processes that can be observed differently when we change the spatial scale. There is recent debate over whether patchy-scale effects are actually outweighed by broad-scale ones, including positive edge effects, reduced competition between species, increased habitat diversity, spread of risks, and higher success in moving between patches. The idea of fragmentation being harmful for biodiversity has been called the "zombie idea" by some researchers, who consider it "an idea that should be dead, but somehow remains alive." It is in fact productive to debate on these core ideas in ecology and conservation, mainly in order to unify opposing views and create better conservation approaches.

Endless Debates about Habitat Fragmentation

The editorial board of *Biological Conservation* has taken into account some empirical evidence that can help in this debate. Many papers have shown through empirical and simulation studies that biodiversity is influenced by the amount of habitat at all scales, and because fragmentation is often correlated with habitat loss, it should also be a target of concern. The small number of papers about fragmentation at landscape level also contribute to these doubts, and therefore we need to improve our understanding on this topic. The scale dependency of population and community dynamics makes it difficult to define thresholds and tipping points for conservation purposes as well.

Another very important point in this debate is the *matrix composition*, which refers to vegetation that is adjacent to the natural fragments, and that may or may not be suitable for determined species. The matrix composition influences species persistence, dispersal ability, and fragmentation. In the case of rainforests, when pastures take their place and lead to a high number of edges in the forest patches, fires are more likely to start, creating a fire-fragmentation tipping point.

In addition, misunderstandings of terminology in fragmentation studies—such as "positive" and "negative" statistically significant effects—when reading scientific papers can also lead policy makers to take wrong actions; such terminology should therefore be reviewed and clarified before being made available to the public.

Finally, the newest views that fragmentation might not be as harmful as we thought are indeed important in the conservation debate scenarios, mainly because they ensure that we keep looking for evidence and answers beyond our previously believed ideas. This discussion is definitely not over; its continuation will greatly help in decision making to better manage the world's fragmented areas.

An Evolutionary Approach to Conserving Plant Habitats

Mackenzie Myers

To conserve plant habitats, a traditional approach to biodiversity—*species richness*, or saving as many species as possible—might not be the most effective route. Instead, vulnerable landscapes might be better served by a quality-over-quantity mindset, a recent paper from a team of UC Berkeley scientists suggests.

Think of going into a grocery store. On a budget and with limited cooking time, shoppers probably don't buy the first dozen random ingredients they see on the shelf. Rather, they find it more practical to shop deliberately, perhaps by looking for ingredients that have a long shelf life, are versatile, and can create a lot of meals. By being intentional about what we buy, we can get more bang for our buck, so to speak. And the same is true for which plant species we conserve.

Place Meets Past

Led by Matthew Kling, a PhD candidate at UC Berkeley's Ackerly Lab, the researchers have applied this sort of thinking to conservation in a recent study examining ways to save plant habitat in California. Using data from extensive plant collections, the team looked at the geography and phylogeny (evolutionary genetic history) of more than 5,200 native California plant species. Kling and his team posit that looking at plants' pasts can reveal which species are likely to thrive and contribute to biodiversity in the future.

Examining the records of 1.2 million individual specimens, the group considered three aspects of how each of the studied species came to be: survival time, or how long its ancestors had been around; divergence, or how many times its ancestors had mutated; and diversification, or how many times its ancestors had branched off into other species. By focusing on these traits, the researchers aimed to select species that are likely to last a long time, adapt to fit a changing environment, and diversify their own future generations.

The team also considered California's landscapes, noting where each species occurred and how intact and well protected each parcel was. For example, a public national park such as Yosemite would be in good condition and protected, while a privately owned beach turned into condominiums would be in poor condition and lack protection. In mapping these aspects, the researchers used gradients, rather than black-and-white, binary definitions, in order to more realistically represent land conditions and the likelihood of plants' survival.

Priorities for Protecting Plant Habitats

The team then created maps of California, including one demonstrating the usual species-richness approach, and drew up a list of the top 100 conservation priority sites in the state. When the maps were adjusted to highlight certain aspects of the data over others, the priority areas tended to shift. But a few locations stood out overall as important to protect: pockets right along the coast, the Coast Ranges, and the Sierra Nevada foothills.

While the study's list of priority areas did provide insight into California's "conservation gaps," some already well-protected areas also made the list. In some cases, the conservation priorities would save a large portion of certain species' growing ranges but a "relatively small tract of land" overall, which could be good news for a rapidly developing, highly populated state that's also home to several rare endemic plant species.

The paper noted a few caveats, such as the possibility that undocumented biodiversity "hotspots" were overlooked, since the team opted to use already recorded data instead of taking field samples. And though it does join a growing wider discussion of how evolution can play a role in conservation, this is the first study to compare these three evolutionary traits alongside traditional species richness, Kling tells Science Connected. An evolutionary, genetic approach to conservation is gaining traction, but scientists are still figuring out which traits are the best to consider in saving wild areas—and "best" may change from one state or region to another, depending on the species present, how well land is protected, and how close the land is to its natural state.

Exploring Possibilities

Kling tells Science Connected that he hopes the study can play a role in saving California landscapes and habitats. The team is in discussions with the Nature Conservancy and the California Native Plant Society to form a statewide initiative for conservation priorities, Kling says, and the study has resulted in a free, public-access tool where users can explore and find patterns in California's conservation gaps (https://bnhm-shiny.berkeley.edu/cappa/).

Whether on a local, regional, national, or international scale, studies like this one may help us figure out a burning question in our rapidly changing world: While using the least amount of resources, how we can best ensure biodiversity for future life on Earth?

Do Humans Influence Jellyfish Coastal Populations?

Laura Treible

Jellyfish are a common nuisance to beachgoers in the summertime, but why certain years have massive jellyfish blooms while others do not is often a mystery. In recent years, some areas appear to have larger and more frequent blooms, so it is important to determine the causes of blooms in general, as well as whether jellyfish blooms are increasing.

How Humans Impact Jellyfish

Human activities such as changing our global climate, fishing, and increasing pollution or nutrients entering waterways influence the oceans in many ways. Some scientists speculate that many of these changes in the oceans that would negatively impact most species would actually result in more jellyfish. My research team and I therefore wanted to look at the effects that humans might be having on jellyfish populations.

One major human-induced problem happening in the oceans is hypoxia, or low oxygen. Hypoxic areas have oxygen values that are too low for many marine animals to survive and breathe normally. Recent work has shown that in addition to the stress of low oxygen, these areas are often accompanied by reduced pH, where the water becomes slightly more acidic. Some evidence suggests that jellyfish can handle the stress of low oxygen or low pH alone, but it is important to consider the combined effects of low oxygen and low pH to reproduce the type of conditions that these organisms naturally experience.

[Editor's note: While a burgeoning jellyfish population is great for the jellies, it isn't great for everyone else. Combined with the low oxygen levels already contributing to population decline, the ever-increasing presence of jellyfish—and their tendency to ingest fish larvae—makes it difficult for other species to survive. Massive jellyfish populations have been known to clog fishing nets, deplete fish populations in farms and in the wild, and even capsize fishing boats. They've clogged power plants and closed

beaches. Understanding how humans have contributed to jellyfish blooms can help us come up with a solution.]

Jellyfish Polyps Reflect Population Health

What most people think of as a jellyfish is actually just one class of many jelly-like animals that live in the ocean, known as a scyphozoan or "true jellyfish." Scyphozoans have unique life cycles that alternate between the adults that we see floating in the water and the microscopic, bottom-dwelling polyps. Polyps reproduce by cloning themselves and producing more polyps, which can provide an easy way to determine how healthy a population is; in more favorable conditions, polyps will reproduce more.

To look at how low oxygen and low pH impact jellyfish, we performed an experiment with moon jellyfish polyps. The polyps were exposed to each factor individually as well as both combined, and we counted the number of new polyps that were produced. We also measured respiration to see if the low oxygen would affect the polyps' metabolism.

After one month, we saw that the polyps were able to survive and reproduce in the low oxygen conditions, but they were reproducing at a slower rate. Surprisingly, lower pH did not affect polyp reproduction, and respiration wasn't affected by either of the conditions. Overall, low oxygen had minimal effects on moon jellyfish polyps, which shows that the jellyfish polyps potentially could survive and still reproduce where other animals might not be able to.

The Jellyfish Advantage

One reason that the moon jellyfish polyps might be able to handle low oxygen or low pH is because of their jelly-like bodies. Jellyfish are made up mostly of water, so they don't need to spend as much energy maintaining basic bodily functions. This allows them to grow quickly without much energy and may help them survive in more stressful conditions. Moon jellyfish are also found along many coasts worldwide, which most likely reflects an ability to live in a wide range of environmental conditions.

Humans are affecting the ocean in a multitude of ways, including by increasing agricultural runoff, which lowers oxygen along many coasts. This lower oxygen and resulting lower pH will have various negative impacts, but jellyfish may be able to handle these stressful conditions better than other marine animals. While it is not confirmed that these changes in the oceans will cause more jellyfish blooms, human activities could be impacting where and when jellyfish bloom.

This study was published in the journal *Marine Ecology Progress Series* and was supported by the National Science Foundation (NSF) East Asia and Pacific Summer Institute (EAPSI) fellowship.

Is Climate Change Causing These Moose to Shrink?

Emily Rhode

The effects of climate change are starting to show themselves in strange and unexpected ways. For the cold-adapted moose of Isle Royale, Michigan, a warming environment could literally be causing them to shrink.

According to researchers at Michigan Technological University, the flourishing moose population at the heart of the world's longest-running predator-prey relationship study is displaying alarming changes that may be most readily explained by environmental pressures. The forthcoming 60-year study of the dynamics between the moose and wolf population of the tiny island in the middle of Lake Superior will show how changes in the numbers of each species have affected the natural ecosystem.

A Dynamic Dance

Moose arrived on the island in the early 1900s and lived unchallenged by predators for over 40 years. When wolves crossed an ice bridge from Canada, the impact of the moose on the natural ecosystem began to change.

In 1958, Durward Allen began the Isle Royale wolf-moose project with hopes that learning more about wolf behavior would stop the slaughter that had brought wolves close to extinction at the hands of humans. Over the next several decades, scientists watched closely as the moose population expanded and then shrank again, in a dramatic dance that was the reverse of the "boom and bust" cycles of the wolf population. The two species were keeping each other in check.

The Shrinking Moose of Isle Royale

This cycle continued until the early 1980s, when parvovirus ravaged the wolf population, bringing their numbers down to only 14. The population never fully recovered, and scientists guessed that this was a result of extreme inbreeding. In the meantime, the moose population climbed once again. Several more cycles of population highs and lows, forced by harsh winters, hot summers, excessive parasites, and the introduction of new genetic diversity to the populations, bring us to the study conducted by the researchers at Michigan Tech.

The team measured the height, width, and length of over 600 moose skulls collected throughout the island by citizen science volunteers. The results showed a 16 percent decrease in the size of the skulls over a 40-year time period. Upon further study, a pattern emerged that showed that moose that live through a warmer first winter tend to grow to be smaller adults and have shorter lifespans.

According to the lead researcher Sarah Hoy, "The conditions you're born into have a massive impact on not only how big you are but also how long you're going to live. This idea isn't new—what we're trying to do is establish how climate warming is affecting this iconic, cold-adapted species."

Compared to populations in places with similar climates, such as Minnesota, the Isle Royale moose numbers are significantly higher. The number of moose in northern Minnesota has shrunk by half over the last 12 years, and scientists believe that climate change has played a big role in the die-off. The spread of parasites, specifically a fatal brain worm that is spread by white-tailed deer, has increased as warmer temperatures allow the deer to move farther north into territory occupied by the moose.

"The moose populations in northern Minnesota have tanked," Hoy says. "Climate is considered a main driver, whether it's direct through warmer winter temperatures causing heat stress and influencing the nutritional condition of moose or indirectly by establishing more favorable habitat for white-tailed deer." Moose on Isle Royale now number over 1,600. That's nearly triple the population from a decade ago, thanks to the number of wolves dropping to only two individuals.

Adaptation or Environmental Pressure?

So what's causing the Isle Royale moose to shrink? Past research has shown evidence that warming climates affect the body size and composition of certain species, but little has been done to study long-lived vertebrates such as moose. While these body composition trends appear to be linked to changing climate, no one is really sure if the changes are a direct negative effect of climate change, or if they represent an animal's ability to adapt to the changes. Smaller body size is typically seen as an advantage in warmer weather because it makes it easier for an animal to regulate its body temperature.

While the study's results suggest that the change in skull size is not a result of adapting to the warmer winters, the Michigan Tech team remain unsure of the exact cause of the observed relationship between warmer winter temperatures and body size, growth, survival, and lifespan of the moose.

Researchers think that it may be due to pressures from an exploding population and a warming climate. In order for the ecosystem to remain balanced, there must be enough resources to sustain all of the individuals. As the population grows without predators to keep it in check, the moose may struggle to find enough food. This competition could lead to malnutrition, stunted growth, and a shortened life expectancy.

An Uncertain Future

The United States National Park Service is currently studying the possibility of introducing more wolves to the island to help keep the moose population in balance. The researchers believe this might be a better outcome for the moose than what is currently happening due to other stressors.

"Decreasing skull size may be an early indicator of population change," says John Vucetich, one of Hoy's collaborators and a professor of ecology at Michigan Tech. "We're likely looking at a population in transition, and the healthiest transition would almost certainly involve restoring wolf predation to Isle Royale."

While many questions about the effects of climate change remain unanswered, researchers will continue to look to the moose and wolf populations of Isle Royale as an ever-evolving case study for the influence of a warming world on predator-prey dynamics.

Plants Can Win a Battle against Aphids

Radhika Desikan

Being sessile, or immobile, plants are faced with constant pressures from their environment, such as extreme climates, microbes, and herbivores including insects and animals. To cope with these challenges, some plants have evolved the ability to tolerate particular stresses or defend themselves against insect pests. For decades, scientists have been trying to understand how some plants tolerate these challenges, with the aim of improving overall plant health and increasing crop yields.

Aphids Are a Problem

Aphids are common pests of most cultivated plants. They belong to a family of sap-sucking insects that feed on plants, causing loss of plant vigor and distorted growth. They are commonly known as greenflies or blackflies, but they can also be other colors. Aphids secrete a sticky substance called honeydew on the surface of plants, which results in the growth of mold, endangering plants' health. Some aphids also transmit viruses that cause plant diseases.

Management of aphid infestations has not been easy for farmers. While aphids have natural predators such as ladybirds (ladybugs) and hoverfly larvae, introducing these is not a guaranteed method to consistently protect crops in the field. Natural pesticides against aphids also exist; they are less harmful to the environment, but their use might not always be economical for farmers. Aphids tend to attack maize near deeper parts of leaf sheaths that are hard to visually detect until the infection has already spread. At this point, farmers quite often resort to using synthetic pesticides which inhibit aphids from reproducing, thereby protecting plants from further attack. The downside of using these synthetic pesticides is that they are harmful to the environment, contaminating water bodies and therefore aquatic life. So, how can this be avoided and how can farmers use better methods to protect plants and the environment?

Plants Have a Solution

Traditional breeding methods have produced several varieties of crops, such as maize, that are naturally resistant to different stresses including aphid attack. The challenge has been to identify visual traits and genetic markers associated with aphid resistance. In addition, identifying cellular changes in a plant specifically associated with aphid resistance would benefit breeding programs. In a new study, Varsani and colleagues have identified mechanisms by which a maize inbred line is naturally resistant to corn leaf aphids (CLA). Their findings highlight previously unidentified cellular processes that lead to resistance to CLA.

Plants that are resistant to aphids exhibit an increase in deposition of callose, a complex compound that is deposited in phloem cells, from which aphids suck the plant sap. Callose deposition leads to blocking of the phloem vessels, thereby inhibiting the food source for aphids and leading to their death. Plants also defend against aphids by producing secondary metabolites and other defense signals such as the hormone jasmonic acid (JA). Using the maize inbred line Mp708, the researchers found that this CLA-resistant line showed an increase in callose deposition in the phloem vessels, compared to another line that was susceptible to CLA. They also found that Mp708 plants had increased levels of a secondary metabolite, 12-oxo-phytodienoic acid (OPDA). OPDA appeared to cause an increase in callose deposition, leading to resistance in Mp708 plants. However, OPDA on its own did not affect the viability of aphids; rather, it acted via biosynthesis of the hormone ethylene, not JA, in the resistant plants but not in the susceptible plants.

By using these findings, scientists can now examine whether selecting and breeding plants that have increased levels of OPDA could make them more resistant to aphids. This would ultimately help farmers who cultivate maize to have better yields and less likelihood of attack from CLA. If it also turns out that OPDA is essential for resistance to different aphids by all plants, the future would hold good prospects for aphid-free plant breeding. We could potentially grow plants without having to worry anymore about these destructive pests in our gardens.

This study was published in the journal *Plant Physiology*.

Do We Really Need Fertilizers To Grow Crops?

Radhika Desikan

We all learn that plants can make their own food via a complex process called photosynthesis. However, to make their food and to grow properly, plants need nutrients—chemicals such as nitrogen, phosphorus, potassium, and calcium—from the soil.

Although these chemicals are naturally present in most soils, years of intensive farming have depleted soils of these nutrients. As a result, humans have resorted to using artificial fertilizers, which are basically synthetic nutrients manufactured through a process that uses a lot of energy and produces a lot of chemicals that are bad for the environment. But, do we really need fertilizers to successfully grow crops?

Of all the nutrients plants need, nitrogen is very important, as it is necessary to form chlorophyll, the green pigment in leaves used for photosynthesis. It is also an important part of DNA and proteins, which are the essence of all life. Unfortunately, nitrogen present in the air as dinitrogen is inactive and useless, and needs to be converted into an alternative form of nitrogen, such as nitrate or ammonium, that can be taken up by plants from the soil. Nitrogen is fixed by special bacteria, *diazotrophs*, that live in the soil, which possess particular enzymes called *nitrogenases*. Some diazotrophs are associated with a family of plants called legumes (including beans, peas, chickpeas, and lentils) that allow nitrogen fixation by the formation of specialized globular structures called root nodules. However, in many cereal plants such as wheat, barley, and maize, nodule-forming diazotrophs are not present; therefore growing these crops heavily depends on the use of nitrogen fertilizers.

Getting the Nitrogen Fix

For several decades, plant scientists have been trying to engineer cereal crops to fix their own nitrogen, meaning that they insert new genes into cereal plants that enable them to fix nitrogen. However, it is also possible that apart from the many cultivated species of cereals grown today, there are indigenous varieties that have the ability to fix their own nitrogen and grow in nitrogen-depleted soils. In search of this, Bennett Van Deynze and his colleagues looked for indigenous varieties of maize in Mexico that could grow in poor soils depleted of nitrogen. They identified a landrace called Sierra Mixe maize near the Oaxaca region of Mexico that predominantly fixed its nitrogen from the atmosphere, not soil.

The scientists observed that this landrace of maize had noticeable aerial roots that grew above the ground at the base of the shoot region. Whilst most varieties of maize have aerial roots that reach the ground to provide more physical support for the plant, in the Sierra Mixe maize an increased number of aerial roots did not reach the ground. Remarkably, these aerial roots secreted a viscous, sugary solution called mucilage. This mucilage was found to contain a number of diazotrophs, which possessed nitrogenase activity. The researchers demonstrated that these bacteria could effectively fix atmospheric nitrogen and transfer it to the maize plant. This is the first report to identify high levels of nitrogen fixation by an indigenous maize plant. Scientists believe that the ability of this plant to fix nitrogen, along with research from other ancient plants, can offer new solutions towards biological nitrogen fixation in cereal crops.

So, it is possible that in the future, farmers might be able to grow maize and other cereals without the need for harmful fertilizers that damage the environment.

Forest Restoration, Not Plantations, Will Curb Global Warming

Neha Jain

Forests are our best natural weapon against climate change. By sucking up large amounts of carbon dioxide from the air, forests can store about a quarter of the carbon necessary to restrict global warming to 1.5 degrees Celsius above pre-industrial levels.

So, it is not surprising that boosting forest area through restoration has been one of the main goals of international organizations tackling rising global temperatures. Encouragingly, 43 countries concentrated around the tropics, where trees grow fast, have pledged to restore 292 million hectares of forest through the Bonn Challenge, a global effort launched in 2011. The goal is to restore 350 million hectares—an area over the size of India—by 2030.

But now a team of scientists from UCL and the University of Edinburgh in the United Kingdom have unveiled a major flaw in the strategies outlined by the countries to restore forests: classifying monoculture plantations (where a single species of tree is planted) as "restoration." In a new analysis of carbon uptake, the researchers highlighted that plantations are far less effective at storing carbon than natural forests are. Consequently, they urged forest scientists and policy makers to change the definition of "forest restoration" to exclude plantations and prioritize natural restoration, recommending several changes to achieve this.

"There is a scandal here," says lead author Simon Lewis in a press release. "To most people forest restoration means bringing back natural forests, but policy makers are calling vast monocultures 'forest restoration.'"

Establishing commercial monoculture plantations is a popular strategy among the countries comprising 45 percent of the area pledged for restoration. But this will seriously undermine efforts to reduce warming, Lewis, Charlotte Wheeler, and colleagues argue, because the amount of carbon stored by plantations, they calculated, is 40 times lower than that sequestered by natural forests.

Natural Forests versus Plantations and Agroforestry

Of the 43 countries—including China, India, and Brazil—that have pledged to restore forests, 24 have already published details of their strategies that cover two-thirds (196 million hectares) of the total pledged area. Three main strategies emerged: leaving degraded agricultural lands to restore into natural forests on their own, converting marginal agricultural lands into plantations of trees such as *Eucalyptus* for paper or *Hevea brasiliensis* for rubber, and growing trees alongside crops (agroforestry).

Alarmingly, almost half of the area—45 percent—committed by the 24 countries was devoted to commercial plantations, followed by 34 percent left to restore naturally, and 21 percent given to agroforestry.

The team compared the amount of carbon sequestered by these strategies in several possible scenarios. If all of the 350 million hectares was left for natural forest restoration, 42 billion tonnes of additional carbon would be removed by 2100. But assuming that all of the other countries that have not yet detailed their plans restore the entire pledged area with the same current proportions of strategies outlined by the 24 countries, then carbon removal would drop to 16 billion tonnes (assuming that the restored forests are protected until 2100).

According to the team, only 7 billion tonnes of carbon would be removed if all of the land was used for agroforestry, and this would further shrink to just 1 billion tonnes if all of the land became monoculture plantations. This means that natural forests are six times better than agroforestry and 40 times better than plantations at removing carbon.

Plantations are poor at storing carbon because they first require land to be cleared, which releases carbon back into the atmosphere. And although fast-growing trees such as *Eucalyptus* and *Acacia* take up carbon, said the researchers, they are usually harvested once every 10 years, releasing the carbon once again into the atmosphere by the decomposition of plantation waste and products such as paper and wood chips.

Defining Forest Restoration and Exploring Solutions

While plantations bring economic benefits, they should not be part of forest restoration, said the team, pressing policy makers to urgently change the definition of "forest restoration" to exclude monoculture plantations. "The UN's Food and Agriculture Organization definition of a forest is extremely broad and includes plantations," explains Lewis.

"Natural forest restoration is clearly the most effective approach for storing carbon," emphasize the researchers. They suggest that countries should increase the amount of land they set aside for natural forest restoration, especially nations located in the humid tropics—regions such as Amazonia, Borneo, or the Congo Basin—because they support high biomass forests compared with drier regions. In some cases, international payments from certain carbon and climate adaptation funds may encourage action.

The scientists recommend that degraded forests and partly wooded areas should be targeted for natural regeneration, whereas treeless regions can be used for plantations and agroforestry systems. But because the calculations show that plantations are the worst option for storing carbon, agroforestry should be prioritized over plantations.

Once restored, natural forests must be protected from future conversion, the researchers stress. They propose a few ways this could be achieved, including "giving title rights to Indigenous peoples who protect forested land, changing the legal definition of how land may be used so it cannot be converted to agriculture, or encouraging commodities companies to commit to not clearing restored natural forests."

This study was published as a comment piece in the journal *Nature*.

Commodity-Driven Deforestation Threatens Forests

Megan Nichols

The global economy is at the mercy of its consumers, whose needs often have a negative impact on the environment. A recently published study explores the impact of commodity-driven deforestation on forests around the world. What is the difference between deforestation and temporary forest loss? What sort of impact is this commodity-driven deforestation having on global ecosystems?

Zero-Deforestation Agreements

The commodity-driven economy is contributing to the decimation of forests around the globe. Projections by the NASA Earth Observatory estimate that if it continues at its current rate, the rainforests could be completely gone within the next 100 years.

Currently, more than 470 companies have made zero-deforestation commitments for many supply lines. The biggest at-risk industries include timber, wood pulp, cattle, and palm oil. While these aren't the only things manufactured in areas that are being deforested, these items and the industries that support them create the largest risk pool. Each company has committed to stopping deforestation in their supply lines by 2020.

From State-Driven to Commodity-Driven Practices

A study published in *Land Use Policy* in 2007 found deforestation in the tropical rainforests started out as a state-driven initiative sometime in the 1970s. Many of the programs started out by encouraging road building and settlement creation in underdeveloped areas. While most of these programs had been phased out by the 1990s, the push for commodity-driven processes has only grown in the meantime. Colonization programs are no longer to blame for deforestation in the tropics. Consumerism is responsible now.

Temporary versus Permanent Deforestation

Satellites can be useful tools for tracking forest health, but they aren't always the most accurate. They are incapable of differentiating between temporary and permanent deforestation.

Global forestry, shifting agriculture, and even wildfires can change the shape of a forest temporarily. Many of the satellite-based tools receive updates only once per year. They would read a forest burned in a wildfire as something lost to deforestation, though the forest itself is likely still standing and will recover over time. Wildfire is the primary cause of deforestation in North America and Russia, for instance.

A team of researchers, including Philip G. Curtis and Christy M. Slay of the University of Arkansas Sustainability Consortium, Nancy L. Harris of the World Resources Institute, and Alexandra Tyukavina and Matthew C. Hansen of the University of Maryland Department of Geographical Sciences, published a study in September 2018. This team found that only 27 percent of deforestation is due to permanent land change. Despite the number of companies making zero-deforestation commitments, the actual deforestation rate hasn't slowed down since 2001.

The research team's data does not include localized events such as windstorms, insect outbreaks, or flooding. These events tend to have a limited impact, and the affected forests will recover over time.

Looking to the Future

Many companies are beginning to look to the future, at least when it comes to agriculture and deforestation. One study found that many international companies have stopped sourcing their beef from cattle farms that have recently cleared forests to create new grazing land. When asked about these agreements, 85 percent of Brazilian ranchers said they were a driving force in their industries.

According to the results the researchers analyzed, this driving force doesn't seem to be enough. The companies trying to create a zero-deforestation supply chain will have to work to save 5 million hectares of forested land every year. (For readers who aren't familiar with the measurement, a hectare is roughly 100 meters by 100 meters of land—the size of two football fields.)

Some governments have started taking steps to lessen the global impact of deforestation. The United Nations launched the Good Growth Partnership to help companies that supply beef, soy, and palm oil to create supply chains that don't rely on cutting down rainforests to meet the growing demand. Experts project demand for these products to double in the next decade, so these changes are more important now than ever.

Changing consumer shopping habits can also help prevent global deforestation. Buying at-risk products from companies that have a completely transparent supply chain will allow consumers to see exactly where their products are coming from. It will also make it easier for customers to see whether a company is living up to its zero-deforestation commitment or not. Coffee, cocoa, and rubber are also products that can contribute to deforestation, though on a smaller scale. Buying from transparent or fair-trade companies can help reduce their impact as well.

Deforestation affects everyone, even if their local forests remain intact. For a more detailed explanation of these findings, look for an article in the September 14, 2019, issue of *Science*.

Living with Wildfires: Fighting Fire with Fire

Nicholas Dove

How do people live with the threat of wildfires, and what can be done to protect lives? That's a question on a lot of people's minds as wildfires get worse and worse each year.

Fires Have Become More Severe

Before we go over how people live with the risk of fires, it is probably worth reviewing how we got here. Many forests throughout the western United States are adapted to frequent, low-severity fire. This means that for millennia, these forests have seen small fires that thin out small trees and dense brush, which reduces the danger of fire for a few decades. This leaves the remaining trees with more sunlight and more fertile soil. Over time, tree species such as ponderosa pine (*Pinus ponderosa*) have evolved thicker, flame-resistant bark, and species including giant sequoia (*Sequoiadendron giganteum*) have evolved cones that only open under intense heat, as in a fire.

While good for the ecosystem, these fires were seen as economically and environmentally destructive by European-descended settlers; during the 1920s, they developed effective techniques to put out a majority of these fires (fire suppression). Fast-forward almost a century, and fire suppression has resulted in forests that are overcrowded and have greater proportions of tree species such as fir and spruce that are better adapted to high-severity fire. This has left many forests and the communities within them extremely vulnerable to the threat of intense wildfires. As a result, recent fires are hotter, bigger, and more destructive. Another reason wildfires in recent decades have become more destructive to humans is that more humans live in fire-prone areas, which are seen desirable places to escape the bustling city life. These areas are called the wildland-urban interface. However, it is important to recognize that humans have lived in these areas for thousands of years.

Living with Wildfires

One of the oldest continuously organized Indigenous communities in the Sierra is the North Fork Mono Tribe. The North Fork Mono Tribe has relied on fire to promote hunting grounds and improve yield of important crops including acorns and pinecones. "Fire [is used] as a tool to take care of the landscape and to maintain our cultural resources, including water," says Ron Goode, Tribal Chairman. When smaller trees are eliminated through fire, the forest sucks up less water from the soil. This water in the soil would otherwise recharge aquifers. In fact, Goode says that "6 percent of California burned before 1860, and 2 percent was burned by Native Americans." In addition to cultivation, the North Fork Mono Tribe have used fire to protect their homes. "You had to be able to put fire out for maybe a mile away, burning so that when fire does come, it's basically a ground fire [that is easier to fight]," says Goode.

The North Fork Mono Tribe are actively promoting the use of fire as a tool, as they have used it for millennia and it contributes to the way they live. For example, the tribe have acted as consultants for the restoration of Crane Valley Meadow. By using fire to remove unwanted plants and restoring the geomorphology of the meadow, they have seen increased beneficial plants such as yarrow and strawberry and greater numbers of deer. Many tribal leaders also sit on the Dinkey Collaborative, a group of stakeholders that work to shape policy for forest management in the Sierra National Forest, sharing their traditional ecological knowledge to improve forest health.

While the North Fork Mono Tribe and other Indigenous Californian tribes have used prescribed fire as a tool for thousands of years, Western society seems to be just catching up. It may seem counterintuitive, but the evidence is clear: a part of the solution in protecting the inhabitants of the Sierra from fire is fire.

How Wildfires Start Their Own Weather

Emily Folk

In the past decade, the United States has seen no shortage of natural disasters. From hurricanes that tear across the coast, destroying homes and flooding properties, to wildfires that consume thousands of acres of land, nature is often vicious and indifferent to human life. But it is also very peculiar. Most consider a wildfire transient in its destruction, a singular event that burns forests and homes before firefighters quell the flames.

But under the right conditions, an intense wildfire can produce its own weather with the potential to cause thunderstorms and, in some cases, "firenadoes"—and the science behind this phenomenon is fascinating.

How Does It Begin?

The atmosphere has to meet specific criteria for a wildfire to create its own weather. The primary influencers are dry air and hot temperatures. According to Amanda Schmidt, AccuWeather multimedia journalist, heat dries out vegetation and makes it more susceptible to the flames, spreading the fire at a faster pace. Hot air creates atmospheric instability, which is ideal for the development of thunderstorms. From here things only continue to worsen.

The fire pushes the air above it upward, and once the air begins traveling, the atmospheric instability accelerates the updraft. As the air continues to rise, the ash gives moisture an opportunity to accumulate and condense into water droplets from which clouds with the scientific designation of pyrocumulus clouds, or "fire clouds," begin to gather and form. If the wildfire is powerful enough and if conditions allow for expansion and aggregation of the clouds, they can grow into pyrocumulonimbus clouds, or "firestorm clouds."

What Are the Effects?

When a wildfire produces its own weather, the effects are both beneficial and detrimental to the environment. In many cases, the clouds only add to the destruction and do little to alleviate it.

While rain may help, lightning does not, as it makes fire management more difficult. As Schmidt states, lightning can strike the already flammable vegetation, creating new fires that place stress on containment efforts. But there are additional dangers just as significant, and one of them comes in the form of shifting wind patterns that fan and spread the flames.

Warm air will rise if its temperature exceeds the temperature of its surroundings, and because this air is lower in density, it can create a vacuum. As the air from around the fire moves into this space, it affects the wind, which in turn exacerbates the problem, because powerful gusts carry the flames and extend their reach to new vulnerable areas.

A combination of wind and wildfires can also result in whirling columns of fire referred to as "firenadoes." These fire swirls carry embers and ash, endangering firefighters who struggle in violent conditions. And the effects of wildfires don't remain isolated to a single place. In some instances, the strength and magnitude of these winds and the sheer heat of wildfires can send particulates and gases all around the world.

How Are Researchers Responding?

In areas susceptible to wildfires, organizations are researching new ways to protect against this destructive natural (and sometimes unnatural) disaster. Studies focus on urban areas to determine what happens when buildings and communities come into contact with a wildfire and how to ensure their safety in the future.

According to Laurel Hamers, a *Science News* reporter, wildfires in an urban area are far more difficult to contain than those that affect only a single house, as the proximity of the buildings to each other often causes them to ignite one another. As a result, after a certain point, curbing the spread of the flames is almost impossible.

Researchers have designed sets of equations that estimate how firebrands (burning vegetation or debris) transfer heat to a surface and how volatile various types of fuel are in different temperatures. Carrying out lab experiments and accumulating data will eventually aid in the development of new protocols and practices.

Further Study and Management

With the many threats a wildfire poses to individuals and communities alike, contending with the danger is no small undertaking. Both research and mitigation are necessary to protect the population from natural disasters, an increasingly relevant responsibility as global warming alters the climate. With the concerted effort of policy advisers and committed organizations, everyone can take steps to reduce the risk of wildfires.

Questions for Discussion

Artificial Night-Lights Are Growing, Getting Brighter

1. What is light pollution, and how does it affect life-forms on Earth?
2. What are some common causes of light pollution?
3. What types of environments are most affected by light pollution, and when?
4. What recommendations would you make for reducing light pollution? What benefits might we expect to see as a result of taking your proposed action?

Globalization and Its Environmental Impact

1. What are some of the pros and cons of globalization?
2. What steps do you recommend that individuals, corporations, and governments take to achieve globalization without sacrificing environmental health?

3. Is "localization," or keeping things in place, a good solution to the negative impacts of globalization?
4. Who is harmed most by globalization? Who benefits most?

Imagining Future Wastewater Solutions

1. What are some sources of water pollution? How can they be mitigated?
2. Describe several methods of remediating wastewater. Which do you think are most viable in your area? Why?
3. Identify a location in the world with water pollution. How would you propose to clean the water and prevent future water pollution in that location?

The Impact of Developing Biofuels on Travel Emissions

1. How are biofuels different from fossil fuels? How are they similar?
2. What are some potentially renewable biofuel sources?
3. Are biofuels the most effective method for reducing travel emissions and traveling sustainably? Why or why not?
4. What policies would you recommend putting in place for transportation fuels and travel emissions locally and nationwide? Why?

Why Plastics Are Dangerous to Our Health

1. What is BPA? Where can you find it?
2. Why is BPA dangerous to human health?
3. How do you recommend avoiding BPAs?

Plastic Pollution: An Emerging Threat Beneath Our Feet

1. What are microplastics, and why are they a global problem?
2. Where do microplastics come from?
3. How would you propose removing microplastics from the environment?

Microbes Help Plants Survive Heavy Metal Stress

1. What is phytoremediation, and how does it work?
2. Do you recommend that farmers use microbes to remediate heavy metal pollution in the soil? If so, how?
3. In what other ways do microbes and plants work together?

Does Habitat Fragmentation Affect Biodiversity?

1. What is the relationship between habitat fragmentation and biodiversity?
2. Do you think that habitat fragmentation is a problem that needs to be addressed? Why or why not?
3. Can you think of any other habitat-related threats to biodiversity? Conversely, what does a habitat need to promote biodiversity?

Do Humans Influence Jellyfish Coastal Populations?

1. What are the causes of increased jellyfish populations?
2. As jellyfish blooms increase, what do you predict will happen to the populations of other marine species?
3. What might prevent jellyfish populations from increasing further?
4. How might humans be able to mitigate increasing jellyfish populations and restore balance to affected marine populations?

Is Climate Change Causing These Moose to Shrink?

1. Why are the Isle Royale moose shrinking?

2. What is the relationship between moose and wolves on Isle Royale?
3. What does this study tell you about the relationship between climate change and animal species?

Plants Can Win a Battle against Aphids

1. Why are aphids a problem for agriculture?
2. Going forward, what do you recommend farmers do to combat aphid infestation in crops? Why?
3. What other traits might it be helpful to develop in crops?
4. Do you think it's dangerous to genetically modify a plant to increase its resilience to pests?

Do We Really Need Fertilizers to Grow Crops?

1. How do fertilizers help with food production? How are they problematic?
2. Describe the role that nitrogen plays in growing plants. Why is it important?
3. Why do some plants have trouble fixing their own nitrogen?

Forest Restoration, Not Plantations, Will Curb Global Warming

1. What is the relationship between natural forests, carbon, and climate change?
2. How do natural forests differ from plantations and agroforestry?
3. This study recommends changing the legal definition of "forest" so that it does not include agriculture. Write a detailed definition of "forest" that you would propose to the United Nations.

Commodity-Driven Deforestation Threatens Forests

1. What is the relationship between commercial agriculture and deforestation?
2. What steps do you recommend that people take to ensure that their shopping does not encourage deforestation?
3. What effects does deforestation have on the surrounding environment? What about on the species that live in forests?

Living with Wildfires: Fighting Fire with Fire

1. What does this author mean by "fighting fire with fire"?

2. Why have wildfires in the Sierra become more destructive to humans in recent years?
3. How do forest fires affect the forest?

How Wildfires Start Their Own Weather

1. How does a wildfire grow into a pyrocumulonimbus cloud?
2. Acting in the role of a policy advisor, what steps would you advise your community to take to reduce the risk of wildfires?

Citizen Science Resources

Kristin Butler, Caroline Nickerson, and Julia Travers

Citizen science is a field as broad as science itself. It can encompass advocacy, policy, education—all undergirded by rigorous scientific standards for data collection and inquiry (Darlene Cavalier and Eric Kennedy Greg Zachary, "The Rightful Place of Science: Citizen Science," 2016). Citizen science is public engagement in scientific research. As Jennifer Shirk, a former president of the Citizen Science Association, said, "Citizen science offers the power of science to everyone, and the power of everyone to science."

Citizen scientists play a game to catch stalled blood vessels to advance Alzheimer's research (the Stall Catchers project); make observations of birds while on a hike (the Audubon Climate Bird Watch); search for extraterrestrial life (SETI@Home)—and even more. SciStarter (SciStarter.org) is the world's largest inventory of citizen science projects, connecting a community of over 100,000 citizen scientists with over 3,000 searchable projects. Projects on SciStarter are searchable by activity (looking for a project to do at home? on a hike?), age group, and other filters.

Citizen scientists in the Boulder Ridge community in Arizona documented the poor air quality in their community through citizen science, using this evidence to push for changes to improve public health. Many citizen science projects are designed for educational outcomes; students in the Broward County public school district in Florida access citizen science projects through a curated portal on SciStarter, thus integrating citizen science with curricular goals, and lifelong learners on iNaturalist are able to investigate the biodiversity around them through the iNaturalist community. As for policy, the future is bright. The Crowdsourcing and Citizen Science Act of 2016 in Congress was proposed to "encourage and increase the use of crowdsourcing and citizen science methods within the Federal Government to advance and accelerate scientific research, literacy, and diplomacy, and for other purposes."

Citizen Science and Climate Change

Some projects, such as ISeeChange, study climate change directly, asking citizen scientists to contribute evidence of a changing climate and unusual weather events. Other projects, including FjordPhyto in Antarctica, study phytoplankton and find evidence of climate change along the way due to increasing environmental disasters affecting all life on earth. Because citizen science allows for more data collection and a larger scope than would be possible with the traditional model of lab-based scientific inquiry, it is uniquely poised to answer the global existential questions posed by climate change.

Climate change and environmental disasters can make individuals feel powerless. Citizen science is a way to take that power back. Citizen scientists help us better understand these phenomena, equipping society to propose climate change solutions. Citizen scientists can use the evidence they collect to push for a better world, learning more about climate, weather, the environment, and natural disasters as they go through the steps of scientific inquiry.

ISeeChange

As Justin Schell said in the intro to SciStarter's podcast episode spotlighting ISeeChange, ISeeChange is "a way to help communities most vulnerable to the effects of climate change document weather in their own backyard and bring greater visibility to these effects through multimedia storytelling." In this podcast, Samantha Harrington, ISeeChange's Digital Community Manager, told SciStarter that one of ISeeChange's mottos is "you are the expert on your own block."

Anyone anywhere in the world can participate in ISeeChange. Simply by making an account and contributing photos and descriptions of weather and climate, these citizen scientists contribute to data used by ISeeChange to draw conclusions about climate change patterns and make recommendations about resiliency, thus making communities more likely to withstand climate change.

Amber Kleinman, one of ISeeChange's first volunteers, also joined the SciStarter podcast. She has been documenting her community in Paonia, Colorado, since 2012, contributing at least one observation a week. As a participant, Amber has been able to study change in her own backyard, as well as explore the website and "bring the scope wider and check out nationwide and worldwide [change]."

Caterpillars Count!

Caterpillars Count! is a citizen science project that conducts surveys of arthropods on shrubs and trees (not just caterpillars, but who can resist a good alliteration?). Anyone anywhere in the world can contribute, either by going to an existing site or creating a new site. Arthropods are an important food source for birds and other wildlife, and studying their abundance allows scientists to assess seasonal change and their impact on bird migration. Dr. Allen Hurlbert, the leader of the project, spoke to SciStarter about the project's relationship to climate action. "There's a gap in knowledge about how the resources birds depend on are responding to climate change, and Caterpillars Count! is addressing that gap."

He also said that Caterpillars Count! was in the process of connecting their data to data from other citizen science projects, such as eBird and iNaturalist, to assess broad impact. According to Dr. Hulbert, Caterpillars Count! is "a great way of getting familiar with the biodiversity around us."

Audubon Climate Bird Watch

The following is an edited excerpt from an article by Julia Travers, previously published on the SciStarter Syndicated Blog Network.

"Hope is the thing with feathers / That perches in the soul," Emily Dickinson wrote. Is there hope for our feathered friends in the era of climate change? Yes, but they need our help. More than 300 North American birds will likely lose over 50 percent of their current range by 2080, according to Audubon's Birds and Climate Change Report. This means the areas with the climate conditions birds need are shifting or disappearing. Like people, birds have to adapt to a changing climate.

For people who want to help birds, Audubon runs a community science program called Climate Watch that gives volunteers resources to monitor range shifts. Learn more about it at https://scistarter.org/audubons-climate-watch.

Within Climate Watch, volunteers team up with coordinators to participate in two annual bird counts across the nation. The data they collect informs Audubon's conservation decisions.

"[Climate Watch] is an opportunity to meet people in their communities and give them the opportunity to connect with nature, to learn about climate change in a way maybe they would not have [done] otherwise," Brooke Bateman, Audubon senior climate scientist, said. She added that people respond to "what they've seen with their own eye."

"I have a lot of hope," said Bateman. "I think the next generation [is] switched on, and that there's [going] to be more solutions ... moving forward. I hope we can keep the momentum going, to really make a difference."

Reef Diving Projects

The following is an edited excerpt from an article in Kristin Butler's Scuba Series, published on the SciStarter Syndicated Blog Network.

This section highlights two citizen science projects that allow divers, many of whom are hobbyists, to contribute to research. Almost 10 years ago, divers Mike Bear and Barbara Lloyd started Ocean Sanctuaries to study sevengill sharks, first off the coast of San Diego and later in other areas, including off the coast of South Africa. Find out more about this project at https://scistarter.org/marine-citizen-science-certification.

Reef Check was founded in 1996. It enlists the help of scuba-diving citizen scientists. Using a rigorous protocol (employing transects and underwater slates), they collect data on coral reefs around the globe and on the rocky reefs and kelp forests off the coast of California. Learn more at https://scistarter.org/reef-check-tropical-ecodiver-training.

Both organizations document important ecological changes and share these scientific findings. Volunteers with Ocean Sanctuaries documented numerous shark sightings off the coast of Southern California in 2014 and 2015, but for the past few years, they haven't documented any. "We are at a loss to explain this kind of thing," said Ocean Sanctuaries co-founder Mike Bear. "There is no shortage of divers in the water on the lookout for this species of shark, but lately we've seen nothing."

Reef Check's research reveals big ecosystem changes along the California coast, said Reef Check California Executive Director Jan Freiwald. Recent Reef Check data shows that several species are being found outside their normal ranges, suggesting that certain species are expanding their historically southerly ranges northward. These species include the crowned sea urchin, rock wrasse, finescale triggerfish, and largemouth blenny. Reef Check data also show that long stretches of kelp forests, especially in Northern California, have disappeared in the last few years due to warming waters and the die-off of sea stars from sea star wasting disease, Freiwald said.

In addition to providing valuable data to understand a changing environment, both nonprofits grow communities of volunteers who care about conservation.

"Volunteering with Ocean Sanctuaries made me realize how dynamic everything is and how connected it all is," said Stefanie Flax. Specifically, she has become more aware of the impact of single-use items, especially plastic. "I'm super aware of what I'm buying now because I've seen where it ends up," Flax said.

For Rick Allen, a volunteer with Reef Check, the biggest personal benefits—in addition to contributing to conservation science—are the friendships he has made through Reef Check. "I came for the data, and I stayed for the relationships," he said.

Start Your Own Project

Many projects relate to the environment and climate change, and they are searchable on the SciStarter project finder, available at https://scistarter.org/finder. If you observe climate change phenomena in your own community and have a unique research question you want to answer, you can create a project of your own on Anecdata.org or CitSci.org, sharing data forms with the citizen science community and inviting contributions. Any project added to these websites is automatically loaded into the SciStarter Project Finder, making it easier for other citizen scientists to find you and help you answer research questions.

Growing by Connecting

In the field of citizen science, we're stronger when we work together. Any of these projects would be happy to work with you, and citizen scientists on SciStarter search for projects every day that they can contribute to. Citizen science is for everyone, and mitigating climate change requires action from all of us.

References

Artificial Night-Lights Are Growing, Getting Brighter

Kyba, C. C. M., Kuester, T., Sánchez de Miguel, A., Baugh, K., Jechow, A., Hölker, F., ... Guanter, L. (2017). Artificially lit surface of Earth at night increasing in radiance and extent. *Science Advances, 3,* e1701528.

Globalization and Its Environmental Impact

Fernando, R. (2012). Sustainable globalization and implications for strategic corporate and national sustainability. *Corporate Governance: The International Journal of Business in Society, 12*(4), 579-589. doi:10.1108/14720701211267883

Haiti's First-Ever Private Nature Reserve Created to Protect Imperiled Species. (n.d.). Retrieved from https://cst.temple.edu/about/news/haiti

Hedges, S. B., Cohen, W. B., Timyan, J., & Yang, Z. (2018). Haiti's biodiversity threatened by nearly complete loss of primary forest. *Proceedings of the National Academy of Sciences of the United States of America, 115*(46), 11850-

11855. doi:10.1073/pnas.1809753115. Abstract retrieved from https://www.pnas.org/content/115/46/11850.

McMichael, A. J. (2013). Globalization, climate change, and human health. *The New England Journal of Medicine, 368*,1335-1343. doi:10.1056/NEJMra1109341.

Imagining Future Wastewater Solutions

Das, P. K. (2018). Phytoremediation and nanoremediation: Emerging techniques for treatment of acid mine drainage water. *Defence Life Science Journal, 3*(2), 190. doi:10.14429/dlsj.3.11346

Hariram, P. (2017). A defining moment for the future of wastewater? Retrieved from http://www.iwa-network.org/a-defining-moment-for-the-future-of-wastewater/

Herold, N. (2018). Water purification and disaster preparedness. Retrieved from https://www.meco.com/water-purification-disaster-preparedness/

Phillips, T. (2019). Bioremediation: Using living organisms to clean the environment. Retrieved from https://www.thebalance.com/cleaning-the-environment-through-bioremediation-375586

The Impact of Developing Biofuels on Travel Emissions

Biodiesel Vehicle Emissions. (n.d.). Retrieved from https://afdc.energy.gov/vehicles/diesels_emissions.html

Biofuels Market Size Will Reach USD 218.7 Billion by 2022, Globally: Zion Market Research. (2018). Retrieved from https://globenewswire.com/news-release/2018/01/09/1285912/0/en/Biofuels-Market-Size-Will-Reach-USD-218-7-Billion-by-2022-Globally-Zion-Market-Research.html

Biofuels: The Benefits and Drawbacks. (2017). Retrieved from https://www.nationalgeographic.com/environment/global-warming/biofuel/

Carrington, D. (2017). Biofuels needed but some more polluting than fossil fuels, report warns. Retrieved from https://www.theguardian.com/environment/2017/jul/14/biofuels-need-to-be-improved-for-battle-against-climate-change

Crutzen, P. J., Mosier, A. R., Smith, K. A., & Winiwarter, W. (2008). N2O release from agro-biofuel production negates global warming reduction by replacing fossil fuels. *Atmospheric Chemistry and Physics, 8(*2*)*, 389-395. doi:10.5194/acp-8-389-2008

Energy Use for Transportation. (2018). Retrieved from https://www.eia.gov/energyexplained/?page=us_energy_transportation

Hill, J., Nelson, E., Tilman, D., Polasky, S., & Tiffany, D. (2006). Environmental, economic, and energetic costs and benefits of biodiesel and ethanol biofuels. *Proceedings of the National Academy of Sciences, 103*(30), 11206-11210. doi:10.1073/pnas.0604600103

Sources of Greenhouse Gas Emissions. (2018). Retrieved from https://www.epa.gov/ghgemissions/sources-greenhouse-gas-emisions

Steer, A. (2015). Biofuels are not a green alternative to fossil fuels. Retrieved from https://www.wri.org/blog/2015/01/biofuels-are-not-green-alternative-fossil-fuels

Sustainable Fuel Sources. (n.d.). Retrieved from https://www.united.com/ual/en/us/fly/company/global-citizenship/environment/sustainable-fuel-sources.html

Why Plastics Are Dangerous to Our Health

Ecology Center. (n.d.) Adverse effects of plastics. *Endocrine Connect, 2*(3), R15-R29.

Lang, I. A., Galloway, T. S., Scarlett, A., Henley, W. E., Depledge, M., Wallace, R. B., & Melzer, D. (2008). Association of urinary bisphenol A concentration with medical disorders and laboratory abnormalities in adults. *JAMA, 300*(11), 1303-1310.

Marques-Pinto, A., & Carvalho, D. (2013). Human infertility: Are endocrine disruptors to blame? Retrieved from https://www.ncbi.nlm.nih.gov/pubmed/23985363

National Institute of Environmental Health Sciences (n.d.). Bisphenol A (BPA). Retrieved from https://www.niehs.nih.gov/health/topics/agents/sya-bpa/index.cfm

Vandenberg, L. N., Chahoud, I., Heindel, J. J., Padmanabhan, V., Paumgartten, F. J. R., & Schoenfelder, G. (2010). Urinary, circulating, and tissue biomonitoring studies indicate widespread exposure to Bisphenol A. *Environmental Health Perspectives*, *118*(8), 1055-1070.

Vandenberg, L. N., Maffini, M. V., Sonnenschein, C., Rubin, B. S., & Soto, A. M. (2009). Bisphenol-A and the great divide: A review of controversies in the field of endocrine disruption. *Endocrine Reviews*, *30*(1), 75-95.

Zoller, R. T., Bansal, R., & Parris, C. (2005). Bisphenol-A, an environmental contaminant that acts as a thyroid hormone receptor antagonist in vitro, increases serum thyroxine, and alters RC3/neurogranin expression in the developing rat brain. *Endocrinology*, *146*(2), 607-12.

Plastic Pollution: An Emerging Threat Beneath Our Feet

Booker, L. (Producer). (2017). *STRAWS* [Motion picture]. United States: By the Brook Productions, LLC.

de Souza Machado, A. A., Kloas, W., Zarfl, C., Hempel, S., & Rillig, M. C. (2018). Microplastics as an emerging threat to terrestrial ecosystems. *Global Change Biology, 24*(4), 1405-1416. https://doi.org/10.1111/gcb.14020

Microbes Help Plants Survive Heavy Metal Stress

Ikram, M., Ali, N., Jan, G., Jan, F. G., Rahman, I. U., Iqbal, A., & Hamayun, M. (2018). IAA producing fungal endophyte Penicillium roqueforti Thom., enhances stress tolerance and nutrients uptake in wheat plants grown on heavy metal contaminated soils. *PLOS ONE, 13*(11). doi:10.1371/journal.pone.0208150

Does Habitat Fragmentation Affect Biodiversity?

Miller-Rushing, A. J., Primack, R. B., Devictor, V., Corlett, R. T., Cumming, G. S., Loyola, R., ... Pejchar, L. (2019). How does habitat fragmentation affect biodiversity? A controversial question at the core of conservation biology. *Biological Conservation, 232,* 271-273.

An Evolutionary Approach to Conserving Plant Habitats

Kling, M. M., Mishler, B. D., Thornhill, A. H., Baldwin, B. G., & Ackerly, D. D. (2019). Facets of phylodiversity: Evolutionary diversification, divergence and survival as conservation targets. *Philosophical Transactions of the Royal Society B: Biological Sciences, 374*(1763), 1-9.

Do Humans Influence Jellyfish Coastal Populations?

Pitt, K. A., Duarte, C. M., Lucas, C. H., Sutherland, K. R., Condon, R. H., Mianzan, H., . . . Uye, S. (2013). Jellyfish body plans provide allometric advantages beyond low carbon content. *PLOS ONE*, 8(8), e72683. https://doi.org/10.1371/journal.pone.0072683.

Treible, L. M., Pitt, K. A., Klein, S. G., & Condon, R. H. (2017). Exposure to elevated pCO2 does not alter reproductive suppression of *Aurelia aurita* jellyfish polyps in low oxygen environments. *Marine Ecology Progress Series*. https://doi.org/10.3354/meps12298

Wallace, R. B., Baumann, H., Grear, J. S., Aller, R. C., & Gobler, C. J. (2014). Coastal ocean acidification: The other eutrophication problem. *Estuarine, Coastal and Shelf Science*, *148*, 1-13. https://doi.org/10.1016/j.ecss.2014.05.027

Is Climate Change Causing These Moose to Shrink?

About the Project: Overview. (n.d.). Retrieved from http://www.isleroyalewolf.org/overview/overview/at_a_glance.html

Hoy, S. R., Peterson, R. O., & Vucetich, J. A. (2018). Climate warming is associated with smaller body size and shorter lifespans in moose near their southern range limit. *Global Change Biology, 24*(6), 2488-2497. doi:10.1111/gcb.14015

Isle Royale National Park. (n.d.). Retrieved from https://parkplanning.nps.gov/parkHome.cfm?parkID=140

Plants Can Win a Battle against Aphids

Varsani, S., Grover, S., Zhou, S., Koch, K. G., Huang, P., Kolomiets, M. V., . . . Louis, J. (2019). 12-oxo-phytodienoic acid acts as a regulator of maize defense against corn leaf aphid. *Plant Physiology, 179*(4), 1402-1415. doi:10.1104/pp.18.01472

Do We Really Need Fertilizers to Grow Crops?

Van Deynze, A., Zamora, P., Delaux, P-M., Heitmann, C., Jayaraman, D., Rajasekar, S., . . . Bennett, A. B. (2018). Nitrogen fixation in a landrace of maize is supported by a mucilage-associated diazotrophic microbiota. *PLOS Biology, 16*(8): e2006352. doi:10.1371/journal.pbio.2006352

Forest Restoration, Not Plantations, Will Curb Global Warming

Lewis., S. L., Wheeler, C. E., Mitchard, E. T. A., & Koch., A. (2019). Restoring natural forests is the best way to remove atmospheric carbon. *Nature, 568,* 25-28. doi:10.1038/d41586-019-01026-8

Commodity-Driven Deforestation Threatens Forests

Curtis, P. G., Slay, C. M., Harris, N. L., Tyukavina, A., & Hansen, M. C. (2018). Classifying drivers of global forest loss. Retrieved from http://science.sciencemag.org/content/361/6407/1108

Gibbs, H. K., Munger, J., Lroe, J., Barreto, P., Pereira, R., Christie, M., . . . Walker, N. F. (2015). Did ranchers and slaughterhouses respond to zero-deforestation agreements in the Brazilian Amazon? *Conservation Letters, 9*(1), 32-42. doi:10.1111/conl.12175

Rudel, T. (2007). Changing agents of deforestation: From state-initiated to enterprise driven processes, 1970–2000. *Land Use Policy, 24*(1), 35-41. doi:10.1016/j.landusepol.2005.11.004.

United Nations Development Programme. (n.d.). Retrieved from https://www.undp.org/

Wallace, P. (2018). Zero-deforestation commodity supply chains by 2020: Are we winning? Retrieved from https://climatefocus.com/publications/zero-deforestation-commodity-supply-chains-2020-are-we-winning

Living with Wildfires: Fighting Fire with Fire

Planning: Dinkey Collaborative. (n.d.). Retrieved from https://www.fs.usda.gov/detailfull/sierra/landmanagement/planning/?cid=stelprdb5351838&width=full

What Effects Tree Thinning Has on Wildfires. (2018). Retrieved from https://www.npr.org/2018/08/07/636423660/what-effects-tree-thinning-has-on-wildfires

How Wildfires Start Their Own Weather

Hamers, L. (2018). Wildfires make their own weather, and that matters for fire management. *Science News Magazine of the Society for Science & the Public, 194*(8), 24. Retrieved from https://www.sciencenews.org/article/wildfires-make-their-own-weather-and-matters-fire-management

Mitton, J. (2018). Wildfires generate their own weather. *Colorado Arts and Sciences Magazine.* Retrieved from https://www.colorado.edu/asmagazine/2018/07/13/wildfires-generate-their-own-weather

Schmidt, A. (2018). How destructive wildfires create their own weather. Retrieved from https://www.accuweather.com/en/weather-news/how-destructive-wildfires-create-their-own-weather/70005643

Citizen Science Resources

Audubon Birds and Climate Change Report. (n.d.). Retrieved from https://climate.audubon.org/

Rare Species Sighted in California, New Species Added to RCCA Survey Protocol. (n.d.). Retrieved from https://reefcheck.org/reef-news/rare-species-sighted-in-california-new-species-added-to-rcca-survey-protocol

About the Authors

Kristin Butler is a freelance journalist and Outreach and Communications Director for the San Francisco Bay Bird Observatory. She holds a B.A. degree in Political Science from Whitman College and an M.S. in Mass Communications from San Jose State University. Kristin has worked as an environmental reporter for The Argus Newspaper and was a monthly contributor to Bay Area Businesswoman News. She's also directed communications and fundraising initiatives for a number of environmental and youth serving organizations, including Earthjustice and Girls Inc. She dedicates her Scuba Series in remembrance of her beloved mother, Marilyn Butler, who passed along to Kristin a deep love for science and nature along with a pair of pink scuba diving fins.

Radhika Desikan is a trained plant scientist and educator. She has researched and published on the behavior of plants facing various abiotic and biotic stresses. Radhika recently became interested in conducting outreach to schools and communicating plant science to a younger audience.

Nicholas Dove is a postdoctoral research associate at Oak Ridge National Laboratory where he studies plant-associated microorganisms and how they impact plant health and ecosystem processes. He has been involved with The Biota Project since 2015, first as treasurer and currently as lead writer. When he is not in the lab or writing for The Biota Project, Nicholas enjoys hiking, skiing, and playing his guitar.

Emily Folk is a sustainability and green tech writer. You can read more of her work on her blog, Conservation Folks at https://conservationfolks.com.

Neha Jain is a science writer based in Hong Kong who has a passion for sharing science with everyone. She writes about biology, conservation, and sustainable living. She has worked in a cancer research lab and facilitated science learning among elementary school children through fun, hands-on experiments. Visit her blog Life Science Exploration to read more of her intriguing posts on unusual creatures and our shared habitat.

Shayna Keyles is the Managing Editor for Science Connected. Since 2016, Shayna has worked with the organization as a way to bridge her passions of understanding the natural world and creating honest, exciting writing. Shayna also serves as an acquisitions editor at North Atlantic Books. Outside of office life, she experiments in the kitchen, doodles furiously, plays board games, and plays outside.

Jacqueline Salvi de Mattos is a plant ecologist and researcher from Brazil. She got into science journalism in order to communicate about the current environmental and ecological crisis.

Caroline Nickerson is the Director of Programs at SciStarter and the Managing Editor of SciStarter's Syndicated Blog Network. Caroline is a Reilly Environmental Policy Scholar at American University, the Executive Assistant of the UF-VA UNESCO Bioethics Unit, and the 2019 Cherry Blossom Princess representing the state of Florida.

Megan Ray Nichols is a science writer and editor of her blog, Schooled by Science. She has a passion for learning and is curious in nature. Her favorite topics to explore include astronomy, the environment, and technology.

Emily Rhode is a science writer and municipal water resources educator. Her goal is to make science accessible and interesting for everyone. She has worked as an outdoor environmental educator, a science teacher, and a professional communicator and trainer. She earned her MEd with a concentration in science education.

Kate Stone is the founder and CEO of Science Connected. She is a social sciences researcher and journalist with 20 years of experience in print and digital media, applied linguistics, and multimedia science communication. She earned her MA from Middlebury and is passionate about science literacy and social good.

Julia Travers writes about science, tech, art, and creative responses to adversity. Her work can be found with NPR, APR, and Earth Island Journal, among other publications. Find her on Twitter @traversjul.

Laura Treible received her PhD from UNC Wilmington in 2018. She is currently an Assistant Professor at Georgia Southern University. Her research interests broadly include gelatinous zooplankton ecology, trophic interactions, coastal ecosystems, and climate change. Specifically, her dissertation research focused on the early life stages of scyphozoan jellyfish (polyps and ephyrae) and their response to various environmental conditions and anthropogenic stressors. Laura enjoys teaching and mentoring students, and communicating science to her students and the public. You can follow her on Twitter @aqua_belle and read more about her research, teaching, and outreach at lauratreible.weebly.com.

About Science Connected

Scientific literacy for all is vital in a modern, high-tech society where we make decisions about important issues such as pollution and climate change, but underprivileged learners face numerous obstacles. Education is expensive. Scientific journals are prohibitively expensive and written at a reading level beyond that of most of the population. Science news in mainstream media is too often reduced to misleading sound bites.

We live in an era of constant scientific discovery and technological change. Science directly impacts our lives and requires our input as informed citizens and voters. The success of nations depends on building a scientifically literate society and a skilled, STEM-educated workforce.

At Science Connected, we believe that widespread science education and responsible science communication go hand in hand. First, we create open, public access to knowledge of the amazing world of science in which we all live. Second, we create equal access to STEM education and careers for women, financially disadvantaged learners, and underrepresented populations.

Science Connected is a 501(c)(3) nonprofit science education publisher. We produce science nonfiction for a general audience as well as affordable resources and labs for science teachers. We are a global team of scientists and science communicators who are dedicated to increasing public understanding of scientific research and creating equal access to science education.

Science Connected gives affordable science lessons and experiments to educators, enables top scientists to share their knowledge directly with the public, and opens doors to science careers for women and underrepresented populations.

This book is part of our ongoing mission to translate complex research findings into accessible science that community members can learn from, discuss, and use to become responsible stewards of the planet we all share. All proceeds from sales of this book support our charitable work and science education outreach programs.

Thank you for joining us in transforming science education and spreading scientific research.

Visit us at www.scienceconnected.org.

More from Science Connected

Think Globally, Garden Locally

This anthology is an investigation into urban gardening, sustainable agriculture, and healthy pollinators. Learn about welcoming pollinators into your garden and growing plants without pesticides. Explore the relationship between chemicals and bee deaths and meet a scientist who became a beekeeper.

Purchase at getbook.at/gardenlocally

Earth Systems: Fun Science Experiments for Children

This book is full of fun science experiments for teachers and parents to enjoy with children ages 5–9 using items commonly found around the home. Explore the physical world; no specialized equipment needed. These experiments are aligned with the Next Generation Science Standards (NGSS).

Purchase at getbook.at/earthsystems

Introduction to Physical Science: A Science Connected Lab Manual

Recommended for children ages 5–9, this book is full of fun science experiments for adults and children to enjoy using items commonly found around the home. No specialized equipment is needed, so jump right in. This guide is aligned with the Next Generation Science Standards (NGSS).

Purchase at getbook.at/physicalsci

Science Connected Magazine

Science Connected journalists and scientists work together to bring you the latest scientific research in a format that everyone can read. We read peer-reviewed journal articles, ask questions, check facts, and write up the results for you.

Read for free at https://magazine.scienceconnected.org

Science Lessons with Science Connected

Bring Science Connected Magazine into your classroom with supplemental texts, discussion guides, and projects for middle school and high school students.

Download this free e-book at
https://www.scienceconnected.org/downloads/lessons

www.ingramcontent.com/pod-product-compliance
Lightning Source LLC
Chambersburg PA
CBHW072143170526
45158CB00004BA/1492